ALL MY FRIENDS ARE PLANETS

THE STORY OF PLUTO

dedicated to future generations of galactic
explorers, astronomers, and lifelong wonderers

look up, blink, and imagine

yours,
alisha vimawala

HI THERE! MY NAME IS PLUTO

I AM THE NINTH PLANET IN THE SOLAR SYSTEM

...OR AT LEAST I THOUGHT I WAS

I HAVE BEEN FLOATING AROUND SPACE

IN A PLACE CALLED THE KUIPER BELT

MILLIONS OF MILES AWAY FROM THE SUN

BUT IT WASN'T UNTIL FEBRUARY 1930
WHEN A CURIOUS 23-YEAR-OLD
NAMED CLYDE TOMBAUGH
FIRST FOUND ME FROM AN OBSERVATORY ON EARTH

IT TOOK A WHILE TO FIND ME

BECAUSE I LIVE SO FAR OUT IN SPACE

THE FURTHEST, IN FACT, FROM THE SUN

WHICH IS THE CENTER OF OUR SOLAR SYSTEM

BEFORE ME,

THERE ARE EIGHT PLANETS

IN THE SOLAR SYSTEM

Mercury

IS THE FIRST AND CLOSEST PLANET
IT CAN ORBIT THE SUN IN ONLY 88 DAYS
VS. THE 365 DAYS IT TAKES FOR EARTH!

VENUS

IS THE HOTTEST PLANET IN THE SOLAR SYSTEM
WHOSE SURFACE HEATS UP TO 900 DEGREES FAHRENHEIT
AND IS ALSO THE BRIGHTEST PLANET IN EARTH'S NIGHT SKY

YOUR HOME, **EARTH**

IS THE ONLY PLANET FOUND TO HAVE LIFE (SO FAR!)
PROBABLY BECAUSE THE MOON STABILIZES ITS ORBIT
MAKING THE WEATHER NICE THERE

MARS

IS KNOWN AFFECTIONATELY AS THE RED PLANET
IN 2015, SCIENTISTS FOUND WATER
SO MAYBE HUMANS MIGHT BE ABLE TO LIVE THERE!

JUPITER

IS THE LARGEST PLANET IN THE SOLAR SYSTEM
MADE ENTIRELY OF POISONOUS GAS

FOR HUNDREDS OF YEARS,
JUPITER ENDURED A STRONG WINDY STORM
WHICH IS WHY IT HAS A GREAT RED SPOT

SATURN

IS MADE ENTIRELY OF GAS, JUST LIKE JUPITER
BUT HAS BEAUTIFUL RINGS AROUND ITS PERIMETER
MADE OF BILLIONS OF SMALL CHUNKS
OF ROCK AND ICE

URANUS

IS ALSO MADE ENTIRELY OF GAS
PARTICULARLY METHANE
WHICH GIVES THE PLANET ITS SPECIAL COLOR

NEPTUNE

IS THE WINDIEST PLANET IN THE SOLAR SYSTEM
WITH WINDS 9X STRONGER THAN ON EARTH
IT TAKES NEPTUNE 165 EARTH YEARS
JUST TO ORBIT THE SUN ONCE!

I WAS THE NINTH

MY BEST FRIEND WAS NEPTUNE

WE WERE THE FURTHEST NEIGHBORS
IN THE SOLAR SYSTEM
AND BECAME FRIENDS INSTANTLY

WE DIDN'T ALWAYS GET TO HANG OUT

ONLY WHEN OUR ORBITS CROSSED

WHICH HAPPENS FOR JUST 20 OUT OF EVERY 248 YEARS

BUT I HAD A SECRET

THAT I COULDN'T TELL ANYONE

NOT EVEN NEPTUNE

I'VE ALWAYS FELT A LITTLE

DIFFERENT

AND IT'S NOT JUST BECAUSE I'M SO FAR AWAY

OR

BECAUSE I'M THE SMALLEST PLANET

OR

BECAUSE I'M ALWAYS COLD

WELL...IT'S ACTUALLY ALL OF THAT

YOU SEE,

ALL MY FRIENDS REVOLVE AROUND THE SUN
IN WHAT THEY CALL AN ORBIT

AND AT THE CENTER OF THEIR ORBITS IS THE SUN

BUT NOT ME!

THE SUN ISN'T AT MY CENTER

I AM ALSO THE ONLY PLANET

TOO SMALL AND LIGHTWEIGHT

TO SWEEP ASTEROID AND OTHER SPACE JUNK

CALLED DEBRIS, OUT OF MY ORBIT

I TRIED NOT TO THINK TOO MUCH

ABOUT THESE DIFFERENCES

NONE OF THE OTHER PLANETS SEEMED TO NOTICE

I WAS STILL THEIR FRIEND ALL THE SAME

...UNTIL MIKE BROWN CAME AROUND

THE DAY WAS AUGUST 24, 2006

MIKE BROWN

AND OTHER ASTRONOMERS FROM EARTH

DECIDED THAT BECAUSE OF MY DIFFERENCES

I WOULD NO LONGER BE CONSIDERED A PLANET

IN THE SOLAR SYSTEM

MY LIFE SUDDENLY BECAME VERY QUIET

ALL THE OTHER PLANETS STOPPED TALKING TO ME

THE ONLY PLANET WHO I THOUGHT

MIGHT UNDERSTAND WAS NEPTUNE

BUT WE WEREN'T GOING TO SEE EACH OTHER

UNTIL THE YEAR 2227

I FELT SO ALONE

EVERYTHING THAT ONCE MADE ME UNIQUE

NOW BECAME A REASON TO EXCLUDE ME

WAS I REALLY THAT DIFFERENT?

WAS THERE ANYONE ELSE LIKE ME?

AND THEN ONE DAY,

I HEARD A FAINT

"HELLO?"

I'D NEVER HEARD THAT VOICE IN MY CORNER

OF SPACE BEFORE!

"WHO'S THERE?" I ASKED.

"I'M MAKEMAKE! (MAH-KEE-MAH-KEE)

I'VE NOTICED YOU IN THE KUIPER BELT BEFORE

BUT I WAS TOO SHY TO INTRODUCE MYSELF."

"...AND I'M ERIS!

WE'RE DWARF PLANETS HERE IN THE KUIPER BELT

JUST LIKE YOU!"

"WHAT'S A DWARF PLANET?" I BLURTED OUT.

"I'VE NEVER HEARD OF THAT BEFORE."

"WE'RE A LOT LIKE PLANETS, BUT SMALLER!" ERIS SAID.

"WE HAVE ENOUGH MASS AND GRAVITY TO BE ROUND AND TRAVEL THROUGH SPACE IN A PATH AROUND THE SUN."

"BUT," MAKEMAKE SAID, "OUR PATH AROUND THE SUN IS FULL OF OTHER OBJECTS, LIKE COMETS AND ASTEROIDS."

"HUH," I MUTTERED TO MYSELF.

THERE ARE OTHERS LIKE ME OUT THERE!

"NEPTUNE," I CALLED,

"MEET MY NEW FRIENDS MAKEMAKE AND ERIS.

THEY ARE DWARF PLANETS JUST LIKE ME!"

"HIYA!

I WAS STARTING TO GET WORRIED, BUT IT LOOKS

LIKE PLUTO HAS FOUND GREAT COMPANY OUT

THERE ON THE BELT!" NEPTUNE REPLIED.

THEY CONTINUED TO TALK AND LAUGH

WHICH MADE ME REALIZE THAT I WAS ACTUALLY

GLAD TO LEARN THAT I WASN'T A REAL PLANET

I WAS DIFFERENT

BUT MY OLD FRIENDS STILL LOVED ME

AND MY NEW FRIENDS WELCOMED ME INTO

THE DWARF PLANET FAMILY

THE BEST PART IS

WE ALL GOT ALONG WITH EACH OTHER

AND TOGETHER

WE MAKE THE SOLAR SYSTEM LARGER

AND FRIENDLIER

WHO ELSE IS OUT THERE?

YOUR TURN! DRAW ME IN!

DRAWING CONTEST

using the blank space on the previous page
draw in your own planet

then snap a picture of your creation
and post it to @storyofpluto on Instagram with
the hashtag #AllMyFriendsArePlanets

for a chance to be featured
or even win a customized plush
of your own drawing!

KEEP LEARNING

www.allmyfriendsareplanets.com

— track the planets in the sky with your smartphone

— see the real photos of Pluto that were transmitted
from the New Horizons space probe

— watch the trailer for Mike Brown's film Planet Nine

— visit your local observatory or planetarium

…and much more!

THANK YOU

to many, many inspiring humans

troy nelson for bringing the solar system to life
sonali jayakar for her sharp writer's eye
ag for encouraging the whimsy
mike brown for killing pluto
(but also opening up the solar system)

and of course,
the vimawala's for being my home
on this beautiful earth